Gabriel Bedinotte e Silva

Verification of adulteration in milk sold in Assis/SP

Gabriel Bedinotte e Silva

Verification of adulteration in milk sold in Assis/SP

Checking for adulteration in pasteurised milk types A, B and C sold in the city of Assis - SP

Imprint

Any brand names and product names mentioned in this book are subject to trademark, brand or patent protection and are trademarks or registered trademarks of their respective holders. The use of brand names, product names, common names, trade names, product descriptions etc. even without a particular marking in this work is in no way to be construed to mean that such names may be regarded as unrestricted in respect of trademark and brand protection legislation and could thus be used by anyone.

Cover image: www.ingimage.com

This book is a translation from the original published under ISBN 978-620-2-19185-2.

Publisher:
Sciencia Scripts
is a trademark of
Dodo Books Indian Ocean Ltd. and OmniScriptum S.R.L publishing group

120 High Road, East Finchley, London, N2 9ED, United Kingdom
Str. Armeneasca 28/1, office 1, Chisinau MD-2012, Republic of Moldova, Europe
Printed at: see last page
ISBN: 978-620-7-98472-5

CONTENTS

CHAPTER 1

INTRODUCTION

Milk is defined as the secretion of the mammary glands of mammalian animals. This product is used as a staple in the human diet in all age groups, mainly because it is one of the most complete products from a nutritional point of view. It has high digestibility, indisputable biological value and is an excellent source of protein and calcium, containing high levels of thiamine, niacin and magnesium (ROCHA, 2009).

Consumers are increasingly concerned about the health and quality of their food. This is prompting the government to define a food safety strategy that encourages producers and entrepreneurs to implement quality control systems that make it possible to trace the food product from its production to its arrival at the consumer's door, i.e. along the production chain (SOUZA, 2006).

This information is no different when it comes to the production and sale of milk, in which we must take all the necessary precautions relating to hygiene, which is directly linked to the health of consumers (SOUZA, 2006).

Even with these laws in force, the consumer is not free from possible adulteration that may occur in the food. The types of fraud are diverse and range from the addition of water to increase the volume of the product, to the addition of substances that, if added in excess or not, may be able to alter the physical and chemical standards of milk, which are established by legislation, such as sodium hydroxide and formaldehyde (SOUZA, 2006).

Nowadays it is recommended by the National Health Surveillance Agency (ANVISA) that whenever we buy any food from a retail outlet, we check its expiry date and whether it has a nutritional table, and this is no different with milk. But even if the milk is not expired and has its nutritional table, this product can be adulterated, which is characterised as fraud, with the addition of substances that alter the chemical or physical properties of the product, which

"""

can even cause illness in the consumer (SOUZA, 2006).

There is also the possibility of increasing the quantity of milk by adding water, all to save on the quantity of this food that will be packaged, consequently generating a much higher profit for the producer, thus deceiving the consumer who pays to buy the milk, and is actually buying milk but added water (SOUZA, 2006).

With this in mind, the purpose of this work was to carry out analyses on samples of pasteurised milk types A, B and C to check that the products were up to standard.

CHAPTER 2

LITERATURE REVIEW

2.1 CHEMICAL COMPOSITION OF MILK

The composition of milk is a determining factor in establishing its nutritional quality and suitability for processing and human consumption. Milk is biosynthesised in the mammary gland under hormonal control. It is estimated that milk has around one hundred thousand different constituents, although most of them have not yet been identified (SILVA, 1997).

The largest component of milk is water, while the rest is made up mainly of fat, proteins and carbohydrates, all of which are synthesised in the mammary gland. There are also small amounts of mineral substances, water-soluble substances transferred directly from the blood plasma, specific blood proteins and traces of enzymes (TRONCO, 2003, p. 17).

The approximate composition of cow's milk is shown in Table 01:

CONSTITUENT	TEOR (g/100 ml)	VARIATION (g/100 ml)
Water	87,3	85,5- 87,7
Lactose	4,6	3,8- 5,3
Proteins	3,25	2,3- 4,4
Fat	3,6	2,4 - 5,5
Mineral substances	0,65	0,53- 0,8
Organic acids	0,18	0,13- 0,22
Others	0,14	-

Table 01: Approximate composition of cow's milk. SOURCE: Adapted from (BEHMER, 1984).

The terms total solids (TS) or total dry extract (TDE) encompass all the components of milk except water. Non-fat solids (NFS) or defatted dry extract (DDE), on the other hand, include all the elements of milk except water and fat (TRONCO, 2003, p. 18).

2.1.1 Colour

The white colour is due to the scattering of light by proteins, fats, phosphates and calcium citrates. The process of homogenising the milk increases the white colour, as the fragmented particles scatter the light more. Skimmed milk has a bluer colour, as there are fewer large particles in the suspension (TRONCO, 2003, p. 18).

2.1.2 Fat

The largest proportion of milk fat is made up of triglycerides (97-98%), small amounts of free fatty acids, sterols and phospholipids. Another advantage of milk fat in human nutrition is the melting point of milk lipids, which occurs below human body temperature (29-32°C) (TRONCO, 2003, p. 18 - 19).

2.1.3 Proteins

Milk proteins are subdivided into casein (80%) and whey proteins (20%). Casein is defined as a complex colloidal substance, associated with calcium and phosphorus, which can be coagulated by the action of acids, rennet and/or alcohol. It is therefore a group of specific phosphoproteins with low solubility at a pH of 4.6 (TRONCO, 2003, p. 19).

Serum proteins, on the other hand. are made up of the following fractions: whey albumin, alpha-lactalbumin, beta-lactoglobulin, immunoglobulins and proteose-peptones. Compared to caseins, whey proteins have little influence on the physicochemical properties of milk. During thermal processing they become important and, at temperatures above 80°C, whey proteins become denatured (TRONCO, 2003, p. 19).

2.1. 4Carbohydrates

Lactose is found in true solution in the aqueous phase of milk. It is a disaccharide made up of glucose and galactose and comes in a proportion of approximately 48g/litre. Lactose is much less sweet than sucrose and the monosaccharides that make it up. When subjected to the heating process, a reaction occurs in the presence of proteins known as the Maillard reaction. This browning reaction is a frequent phenomenon in evaporated and sterilised milks (TRONCO, 2003, p. 20).

Figure 01 shows the structural formula of lactose.

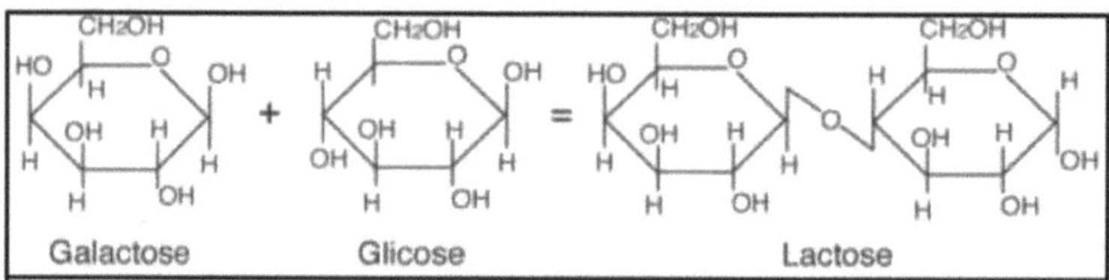

Figure 01: Structural formula of lactose. SOURCE: (TRONCO. 2003, p.20).

2.1.5 Minerals

Mineral substances account for approximately 0.6 to 0.8% of the weight of milk and, in analyses, are referred to as ash, representing the residue that remains after the milk has been subjected to the incineration process. Among the minerals in milk, calcium plays an important role in human health. Calcium and phosphorus are bound to casein in the form of a calcium phosphocaseinate complex. The calcium content of milk is 125 mg/100mL. There are also various minerals in milk, but in very small quantities: sodium, potassium, magnesium, fluorine, iodine, sulphur, copper, zinc, iron, etc. Mineral salts are important and govern the thermostability of milk, as well as coagulation processes (calcium) (TRONCO, 2003, p. 20).

2.1.6 Vitamins

Milk also contains various vitamins with considerable levels of fat-soluble vitamins (A, D, E and K) and water-soluble vitamins (B and C), all of which are susceptible to destruction by various factors: heat treatment, the action of light, oxidisation, etc. For these reasons, when vitamins are added to milk, it is essential to establish adequate control of the amount of vitamins that remain in the milk after treatment (TRONCO, 2003, p. 21). Table 02 shows the concentrations of vitamins in milk:

VITAMIN	TEOR (100 g of milk)
Vitamin A (International Units - IU)	160- 225
Thiamine (Vitamin B) - (Micrograms)	40- 65
Ascorbic Acid (Vitamin C) - (Milligrams)	2,1 - 2,2
Vitamin D (International Units - IU)	1,7
Riboflavin (Vitamin G) - (Micrograms)	195- 240
Nicotinic acid (Milligrams)	2 - 8

Table 02: Amount of vitamins in milk. SOURCE: (BEHMER, 1984).

2.2 MILK PRODUCTION IN BRAZIL

Table 03 shows the number of milked cows and milk production, total and percentage change, according to the Federative Units between 2010 and 2011.

Federation Units	Number of milked cows (1,000 head)			Milk production (1 000 000 litres)		
	2010	2011	Variation (%)	2010	2011	Variation (%)
Total	22 925	23 227	1.3	30 715	32 091	4.5
Minas Gerais	5 447	5 631	3.4	8 388	8 756	4.4
Goiás	2 490	2 616	5.5	3 194	3 482	9.0
Bahia	2 212	2 104	(-) 4.9	1 239	1 181	(-) 4.6
Paraná	1 550	1 589	2.5	3 596	3 819	6.2
Rio Grande do Sul	1 496	1 530	2,3	3 634	3 879	6.8
São Paulo	1 488	1 453	(-) 2.4	1 606	1 601	(-) 0.3
Santa Catarina	979	1 022	4.3	2 381	2 531	6.3
Rondônia	1 083	990	(-" 8.6	803	707	(-) 12.0
Pará	764	795	4.2	564	591	4.7
Mato Grosso	618	634	2.6	708	743	4.9
Pernambuco	576	620	7.6	877	953	8.6
Maranhão	574	592	3.1	376	387	2.9
Ceará	539	550	2.0	444	456	2.6
Mato Grosso do Sul	528	530	0.5	511	522	2,1
Rio de Janeiro	415	427	3.0	489	500	2,2
Tocantins	526	425	(-) 19.1	269	267	(-> 0.8
Holy Spirit	395	409	3.6	437	451	3.2
Rio Grande do Norte	258	262	1,7	229	243	6.0
Paraíba	239	259	8,3	217	237	9.3
Sergipe	221	227	2,7	297	316	6.5
Piauí	158	156	<-" 1,0	87	89	2,0
Alagoas	149	155	3.7	231	238	3,0
Amazonas	112	127	13.1	47	52	10,2
Acre	71	71	1,0	41	42	2.9
Roraima	19	23	18,8	6	7	17,8
Federal District	21	20	<-) 4.9	36	30	(-) 17.3
Amapá	9	11	30.4	7	9	36.4

Table 03: Number of milked cows and milk production, total and percentage change, by Federative Unit between 2010 and 2011.

National milk production increased by 4.5 per cent between 2011 and 2010. Minas Gerais stood out, with a 27.3 per cent share of production, followed by Rio Grande do Sul (12.1 per cent), Paraná (11.9 per cent) and Goiás (10.9 per cent). These states account for 62.1% of all milk produced in the country. Comparing 2011 and 2010, it is worth mentioning the growth in cow's milk production in the states of Goiás (9.0%), Rio Grande do Sul (6.8%), Paraná (6.2%) and Minas Gerais (4.4%), as well as the reductions of 12.0% in Rondônia and 4.6% in Bahia, in addition to the relative stability of production in São Paulo (IBGE, 2011).Figure 02 shows milk productivity according to the 10 municipalities with the highest productivity in Brazil in 2011.

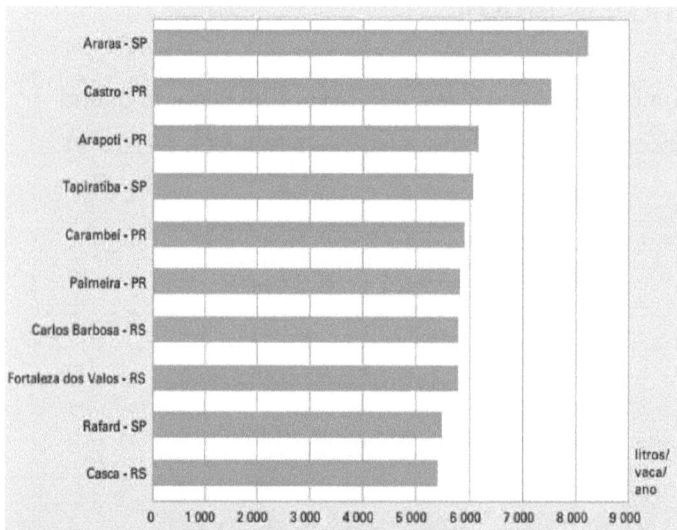

Figure 02: Milk productivity, according to the 10 municipalities with the highest productivity - Brazil - 2011
highest productivity - Brazil - 2011. SOURCE: IBGE, 2011.

The ten municipalities with the highest milk productivity produced over 5,000 litres/cow/year, which is similar to the average in European countries and the United States, and higher than in China and India. Araras (SP) was the municipality with the highest productivity, 8,213 litres/cow/year, although its total production (16.4 million litres) is not significant in the national total. Castro (PR) is the main municipality in milk production and took second place in productivity, with 7,527 litres/cow/year. The main municipalities in terms of milk productivity are located in the states of São Paulo, Paraná and Rio Grande do Sul and represent high-tech, professionalised dairy farming, with selected dairy herds, combined with favourable climatic conditions (IBGE, 2011).According to IBGE data, the municipality of Assis saw a considerable increase in cow's milk production between 2004 and 2007, but since then its production has been falling. It is worth remembering that the amount of cow's milk production means the total amount of milk, in litres, produced during the reference year of the survey by the cows milked in the municipality. Table 04 shows the production figures for cow's milk in the municipality of Assis between 2004 and 2011.

Year	Quantity of Cow's Milk Production (thousand litres)
2011	3200
2010	3300
2009	3430
2008	3700
2007	3900
2006	2800
2005	2700
2004	2780

Table 04: Cow's Milk Production in the Municipality of Assis (2004-2011). Source: IBGE, 2013.

Table 05 shows milk production in the Assis region in 2011, based on IBGE data.

City	Quantity of Cow's Milk Production (thousand litres)
Assisi	3200
Cândido Mota	1443
Cruzalia	978
Echaporã	3427
Florínea	93
Ibirarema	558
Maracaí	935
Palmital	1789

Pedrinhas Paulista	739
Platinum	490
Tarumã	170

Table 05: Cow's milk production in the Assis region - 2011.
SOURCE: (IBGE, 2013).

The city that produced the most milk in 2011 was Echaporã (3,427 litres), followed by Assis (3,200 litres). These cities account for almost half of the region's milk production. The towns of Florínea and Tarumã had the lowest milk production, with 93,000 and 170,000 litres respectively.

2.3 MILK CLASSIFICATION

When we buy milk for consumption, we find it as pasteurised (sachet), UHT (carton), also known as long-life milk, or powdered milk, which is obtained by dehydrating cow's milk. According to the Technical Regulation for the Identity and Quality of Pasteurised Milk (IN 51/02) issued by the MAPA, pasteurised milk is fluid milk made from raw milk that has been refrigerated on the farm and which

presents the production, collection and quality specifications of this raw material contained in its own Technical Regulations and has been transported in bulk to the processing establishment (PAIVA, 2007).

2.3.1 Pasteurised milk

In view of the health risks inherent in drinking milk that has been obtained and processed under unsatisfactory conditions, it is necessary to apply an efficient heat treatment to destroy micro-organisms that does not produce significant alterations to the nutritional quality of the product, such as fat, protein or carbohydrate degradation. It was then found that pasteurising milk can meet these objectives, but it should be noted that pasteurised food must be consumed in a short space of time and that the efficiency of destroying bacteria can vary according to the initial microbial load of the product (ROCHA, 2009). The effect of heat treatment alters the nutrient content of any food, especially water-soluble vitamins. Pasteurisation reduces the vitamin content of cow's milk by 12%, and the main consequences of a pasteurisation process are physicochemical changes, the Maillard reaction, denaturation and coagulation. Long-life milk suffers greater protein denaturation than pasteurised milk (BEUX, 2012).

Pasteurised milk has three classifications: "A", "B" and "C". Type A milk is characterised by a mechanical, closed-circuit milking system with standardised milking parlours. It is immediately refrigerated and the pasteurisation and packaging process takes place entirely on the farm. This process guarantees minimal risk of human contact among pasteurised milks and is only viable for large-scale production. Type B milk differs from type A in that it is collected on the farm in tanker trucks and will be pasteurised and packaged in an industry. The characteristics of the milking parlour are the same as for type A milk. Type C milk can be milked by hand or mechanically. The milk is stored in refrigerated tanks before going to the dairy where it is pasteurised and packaged (SCALCO, 2009, p. 01).

There are two most commonly used pasteurisation processes: slow pasteurisation and rapid pasteurisation.

2.3.1.1 Slow pasteurisation

Brazilian legislation does not allow the use of slow pasteurisation to process milk "for consumption", which is restricted to the processing of by-products, mainly cheese (TRONCO, 2003, p. 60).

The advantage of this system is that it preserves the milk's properties as closely as possible to its *fresh* state. The colour and taste remain unchanged.

However, it does have some disadvantages. As it is a discontinuous process, it requires time, a large amount of heat and cold, as well as a lot of space for its equipment, factors that make it quite expensive. It can also allow thermophiles to develop. This is why it is necessary to avoid the formation of foam, which favours the emergence of thermoresistant micro-organisms. This system is practically out of use in industry, due to economic and technological aspects (TRONCO, 2003, p. 61).

It consists of heating the milk in a cylindrical-vertical, double-walled tank fitted with an agitator. The milk is heated, with constant agitation, to 65°C and kept at this temperature for thirty minutes. The milk is heated by circulating hot water through the double walls of the apparatus. It is then cooled to a temperature of 4 - 5° C by circulating cold water through the double walls of the machine (GUIMARÃES, 2001). Figure 03 shows a slow pasteuriser.

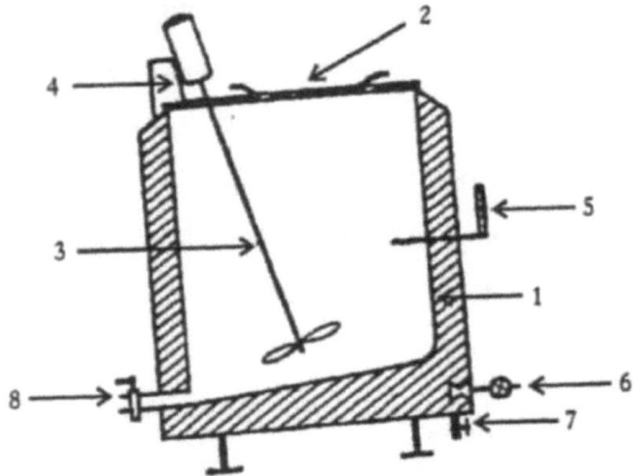

Figure 03: Slow pasteuriser. SOURCE: (TRONCO, 2003, p. 60).

Where,

1. Heating jacket
2. movable lid
3. rotary shaker
4. Agitator support
5. Thermometer
6. vapour injection

7. Purge

8. pasteurised milk output

2.3.1.2 Rapid pasteurisation

Milk pasteurisation temperatures are sufficient to destroy, in addition to pathogenic organisms, all yeasts, all fungi, all gram-negative bacteria and some gram-positive bacteria. In addition to destroying

during pasteurisation there is partial or total denaturation of enzymes, vitamins, denaturation of whey proteins (10-20%), insubilisation of salts, among other effects (TRONCO, 2003, p. 65).

It consists of heating the milk in a cylindrical-vertical, double-walled tank fitted with an agitator. The equipment used consists of a set of plates, all made of stainless steel. The milk is heated and cooled by circulating between the plates, in very thin layers, in a closed circuit, protected from air and light under pressure, at a heating temperature of 72 to 75°C, for 15 seconds and cooled with iced water at a temperature of 2 - 3°C (GUIMARÃES, 2001). Figure 04 shows a set of plates in a rapid pasteuriser.

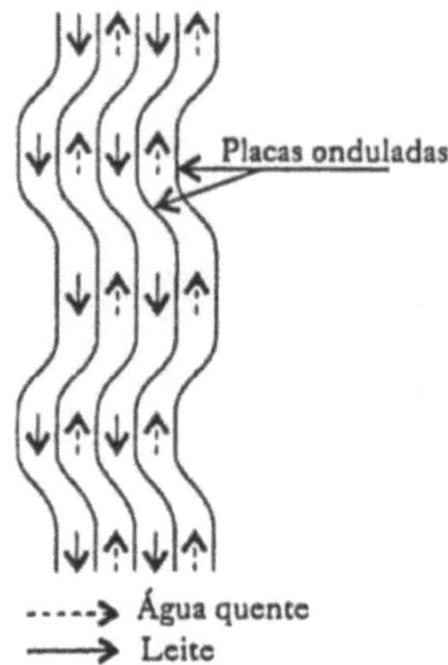

Figure 04: Rapid pasteuriser plate set.

SOURCE: (TRONCO, p. 61).

2.3. 2UHT milk

UHT (Ultra High Temperature) milk means High Pasteurisation Temperature. The milk is homogenised and subjected to a temperature of between 130°C and 150°C for 2 to 4 seconds by means of a continuous flow thermal process, immediately cooled to a temperature of less than 32°C and filled under aseptic conditions in sterile, hermetically sealed packaging (TURCO et al, 2002).

The result is the destruction of all micro-organisms with a quality end product and a shelf life of up to 180 days at room temperature. Another advantage is that the air is removed from the packaging, preventing the fats from oxidising (TRONCO, 2003, p. 66 - 67).

The aim of UHT is to obtain a bacteriologically sterile product that retains the nutritional and organoleptic characteristics of the fresh product, which is commonly referred to as long-life milk (TRONCO, 2003, p. 66).

In UHT milk, 100% of the whey proteins are denatured (TRONCO, 2003, p. 65).

2. 4MILK FRAUD

The RIISPOA (Regulations for the Industrial and Sanitary Inspection of Products of Animal Origin) establishes that the product "milk" cannot be added with substances that are not permitted, characterising its intentional adulteration as fraud, which harms consumers, farmers and the fraudster's competitors (BRASIL, 1952).

Counterfeiting is defined as the partial or total addition or subtraction of any substance in the composition of a product, under such conditions that it does not correspond to the normal product. Checking for counterfeiting in industrial products is relatively simple. However, the same is not true,

with "natural" products, where the inconstancy of the composition must be taken into account, with variations depending on racial, individual, dietary and climatic conditions at the time of analysis (BEHMER, 1999).

The quality of milk is monitored by public health institutes through specific tests involving the determination of density, fat content, rancidity and acidity, and the presence of additives used for preservation or even the identification of foreign materials in the milk to hide its 'baptism' with water. Table 06 shows some of the materials that have been found in milk and their role in the fraud process (LISBÔA, 1997).

MATERIAL	FUNCTION
Antibiotics	To preserve the milk, avoiding the action of micro-organisms.
Urine	**"Disguise" the addition of water to the milk while maintaining the** initial density.
Starch	**"Disguise" the addition of water to the milk while maintaining the** initial density.
Formaldehyde	To preserve the milk, avoiding the action of micro-organisms.
Sodium bicarbonate	**"Disguise" the increased acidity of milk observed** when it is in a state of deterioration.
Boric acid and borates	To preserve the milk, avoiding the action of micro-organisms.
Salicylic acid and salicylates	To preserve the milk, avoiding the action of micro-organisms.

Water	It aims to increase the volume of milk.

Table 06: Materials found in milk and their subsequent function.
SOURCE: LISBÔA, 1997.

Milk is considered to be fraudulent, adulterated or falsified if: it has water added to it, any of its components have been subtracted, except for fat in the "C" and "skimmed" types, preservatives or any elements foreign to its composition have been added to it, it is

of one type and labelled as another of a higher category, if it is raw and sold as pasteurised or if it is exposed to consumption without proper guarantees of inviolability (BRASIL, 1952).

According to TURCO et al. (2002), some factors make monitoring difficult:

- There are few government-accredited analysis laboratories;
- More laboratory technicians need to be trained to carry out the analyses;
- The need for more human and material resources for the inspection process and/or to redesign the process in order to facilitate the work of the current team of inspectors;
- Legal loopholes that allow companies to challenge the results of analyses;
- Relatively cheap punitive fine;
- Possible difficulties for inspectors in working with companies;
- Cost of analyses;
- Excessive focus on quantitative methods.

In order to reduce these problems, some corrective actions are suggested, such as:

- Accrediting new laboratories to carry out both chromatographic and spectrophotometric analyses;

- Find out the ideal number of technicians to meet the demand forecast by the government and then develop an appropriate incentive policy;

D Discuss with public authorities the need for more staff and a possible increase in the budget for inspection bodies. Also reduce the duties of inspectors, keeping the focus on the most critical activities;

- Improving the sample collection process, avoiding the freezing and mixing of samples, as well as avoiding delays in reaching the laboratory, in order to minimise doubts about test results;

- Increase the amount of the punitive fine;

- Increase the rotation of inspectors working in each company, preventing the same inspector from always working in the same companies;

- Encourage spectrophotometry as an alternative method to chromatography;

- Use a combination of methods to complement quantitative analyses. To do this, use tax and process audits as complementary practices.

- In parallel with intensifying analyses of protein content, the current inspection system for detecting non-compliance of fluid milk due to the addition of whey could be improved.

2.5 LEGISLATION

In recent years, the Federal Inspection Service (SIF) has been striving to modernise and improve legislation with the aim of finding alternatives that make Brazilian establishments competitive on the international market and that guarantee consumers on the domestic market the consumption of safe food (TRONCO, 2003, p. 26).

On 8 December 1999, the Department for the Inspection of Products of Animal Origin (DIPOA), whose mission is the hygienic, sanitary and technological inspection of these products, drew up and sent out for public consultation the Basic Technical Regulations for the Identity and Quality of types "A", "B" and

"C" milk, pasteurised milk, refrigerated raw milk, goat's milk, the collection of refrigerated raw milk on rural properties and its transport in bulk. Until legislation was passed, such as Normative Instruction no. 51 of 18 September 2002, published in the DOU of 20 September 2002, section 1, pages 13 to 22 (TRONCO, 2003, p. 27-28).

According to Normative Instruction 51 of 18 September 2002 (BRASIL, 2002), the standards for density at 15°C, °Dornic acidity, Cryoscopic Index, pH, lacto sedimentation and fat percentage are as follows:

It is worth remembering that the type A, B and C milks used in this work were all whole milk. Table 07 shows the physical and chemical requirements for milk.

Requirements	Integral	Standardised	Semi-skimmed	Skimmed
Fat (g/100g)	Original content	Min. 3,0	6 a 2,9	Max. 0.5
Relative density at 15°C (g/ml)	1,027 a 1,034			
PH	6,6-6,8			
Lacto sedimentation	Good			
Acidity (°Dornic)	14-18°Dornic			
Maximum cryoscopic index	- 0.530°H (-0.512°C)			

Table 07: Physical and chemical requirements in milk. SOURCE: Adapted from BRASIL, 2002.

2.6 PHYSICO-CHEMICAL ANALYSES OF MILK

2.6.1 Titratable acidity

The alizarol and alcohol tests are quick platform tests and give an approximate idea of the acidity of milk. But when you need to know the acidity exactly, you should titrate with sodium hydroxide N/9 or 0.11 N, known as Dornic soda. This test is usually carried out in the laboratory (TRONCO, 2003, p. 107).

In the titratable acidity test, a basic (i.e. alkaline) substance, sodium hydroxide (NaOH), is used to neutralise the acid in the milk. An indicator substance (phenolphthalein) is used to show how much alkali was needed to neutralise the acid in the milk. The indicator remains colourless when mixed with an acidic substance, but turns pink in an alkaline medium. Therefore, the alkali (NaOH N/9) is added to the milk until the milk turns pink. Each 0.1 mL of NaOH N/9 solution used in the test corresponds to 1° D or 0.1g of lactic acid/L. Table 08 shows some examples of the results in degrees Dornic, the correspondence in pH and the interpretation with regard to the acidity and heat resistance of the milk (BRITO, et al. 2006).

PH	Dornic acidity °D	Interpreting the results
6,6-6,8	14-18	Normal (fresh) milk.
>6,9	< 14	Typical alkaline milk: milk from cows with mastitis, milk from the end of lactation, retention milk, water-frauded milk.
6,5-6,6	19-20	Slightly acidic milk: milk from the beginning of lactation, milk with colostrum, milk at the beginning of the fermentation process.
6,4	+/- 20	Milk that cannot withstand heating to 110 °C.
6,3	22	Milk that cannot withstand heating to 100 °C.
6,1	>24	Milk that cannot withstand pasteurisation at 72°C.
5,2	55 - 60	Milk that begins to flocculate at room temperature.

**Table 08: Interpretation of pH and
milk acidity results
. SOURCE: BRITO, et al. 2006.**

The ability to combine alkali (NaOH) and fresh milk is provided by the components: salts, milk proteins and dissolved carbon dioxide, as shown in table 09 below.

COMPONENT	% CONTRIBUTION
Casein	0,05-0,08
Citrates	0,01
CO_2	0,01 -0,02
Albumin (whey protein)	0,01
Remaining phosphates	0,06 - 0,12

**Table 09 - Contribution of components to the acidity of
fresh milk
. SOURCE: TRONCO, 2003, p. 109.**

2.6.   2Density

The density of a liquid or solid body is the relationship between the mass (expressed by weight) and the volume of this body. The unit of density is g/cm³ or g/mL (TRONCO, 2003, p. 112).

Density = mass / volume

The density of milk is relative, i.e. the quotient resulting from dividing the mass of a volume of milk by an equal volume of water at a certain temperature. The determination of this parameter serves to control, up to certain limits, fraud in milk, in terms of prior skimming or the addition of water (TRONCO, 2003, p. 112).Table 10 shows the density of the following milk components.

COMPONENT	Density (g/mL or g/cm)³
Water	1,000
Fat	0,930
Lactose	1,666
Proteins	1,346
Minerals	5,500

Table 10: Density of milk components.

The average density of milk can vary from 1.027 to 1.034 g/cm³ (you can also use the expression 27 to 34 °GL - degrees lactodensimetres), decreasing as the amount of fat increases, which happens when you increase the proportion of protein, lactose and mineral salts (TRONCO, 2003, p. 113).

In the routine for determining density, the thermolactodensimeter (Quevene) is used, which has graduations from 15 to 45 °C and a density of 1.015 to 1.045 (g/cm³), a correction must be made when the milk has a temperature different from this. The sample temperature should be below 30°C, preferably between 10 and 20°C. Fat is the only constituent with a lower density than water, so it has the greatest influence on reducing the density of milk (TRONCO, 2003, p. 113).

The density test can be useful in detecting milk adulteration, since the addition of water causes a decrease in density, while the removal of fat results in an increase in density (SANTOS; FONSECA, 2007, p. 314).

2.6.3 Lacto sedimentation

Its purpose is to verify the presence of dirt in the milk from processes prior to its manufacture, such as milking, utensils, transport and others, in order to estimate the hygienic and sanitary conditions of the process.

2.6.4 Cryoscopy

2.6.4.1 Hortvet scale (H)

Julius Hortvet pioneered the use of freezing point as a qualitative analysis of milk. In his studies at the beginning of the century, he came to the conclusion that 7% and 10% sucrose froze at -0.422 and -0.621 degrees Celsius respectively. Later studies measured these temperatures with greater precision, arriving at values of -0.408 and -0.600 degrees Celsius. However, by this time Hortvet's values had already created a standard for milk cryoscopy and were adopted internationally. Fortunately, this didn't cause any problems, since measurements of the amount of water added to milk are relative, i.e. if the scale used is the same (Celsius, Hortvet or any other) the values will always be constant (LAKTRON, 38 p.).

2.6.4.2 Freezing temperature x water percentage

Hortvet based this on the fact that the freezing point of pure water is 0 H and the average freezing point of pure milk is close to - 0.540 H (LAKTRON, 38 p.). Thus, by changing the freezing point of milk, the amount of water added to the milk can be determined. Figure 05 shows the curve of Hortvet values versus percentage of water.

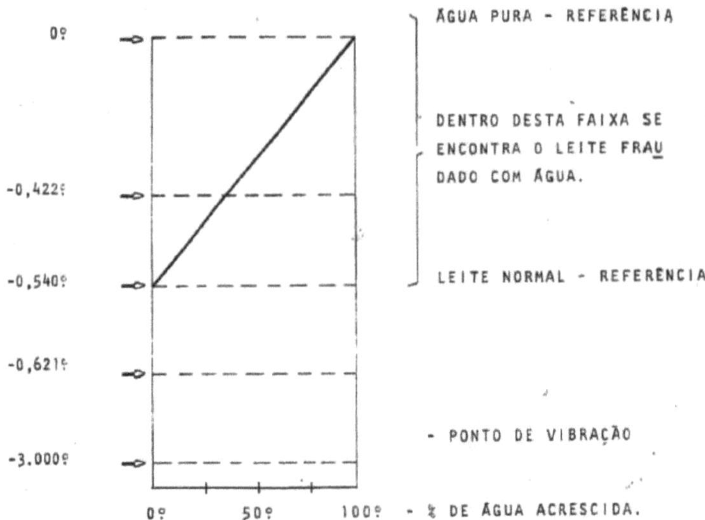

Figure 05: Hortvet values x water percentage curve.
The vertical axis represents the Hortvet scale where you can see the points for pure water (0 H), pure milk (- 0.540 H), maximum cooling (- 3.000 H) and calibration (- 0.422 H and - 0.621 H). We can also see the ranges for milk with water (0 H to - 0.540 H) and calibration (- 0.422 H to - 0.621 H) where the reading is linear. The horizontal axis represents the percentage of water added to the milk. The bold line is the curve that relates the freezing temperature to the percentage of water added (LAKTRON, 38 p.). Cryoscope equipment normally has a linear and precise sensor response curve within the limits of - 0.422 H and - 0.621 H for which the equipment has been calibrated and within which we normally find the results of the milk analysed. Outside this range, the results may show normal deviations because they are outside the calibrated scale (LAKTRON, 38 p.).

2.6.4.3 Freezing point

This is the temperature at which a solution changes from a liquid to a solid state. The fastest way to reach the freezing point of a solution is to cool the solution a few degrees below its freezing point and then apply a rapid mechanical vibration. Normally in a digital cryoscope this vibration is applied automatically when the temperature of the solution reaches - 3,000 H, forming crystallised ices. After vibration, this process generates a thermal imbalance causing the solution to release the heat of fusion, where it passes from the solid to the liquid state, which will cause its temperature to rise until it reaches freezing temperature. The solution will then remain at this temperature for a certain time. Figure 06 shows the "Plateau" (LAKTRON, 38 p.)

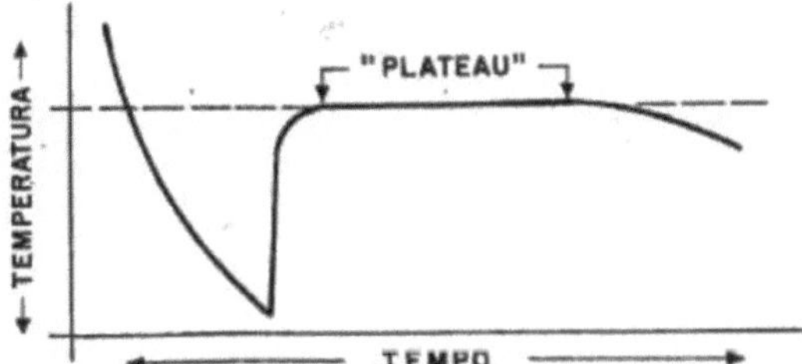

Figure 06: Plateau. SOURCE: (LAKTRON, 38 p.).

This so-called "Plateau" time in the digital cryoscope refers to a time of 12 seconds when the measurement is taken. Therefore, the temperature at which the "Plateau" occurs is a function of the solute and solvent concentration of the sample (LAKTRON, 38 p.).

To give uniformity and speed to the analysis, the test tube used in the cryoscope is immersed in the cooling bath at approximately -7 °C. Variations in the temperature of the bath may vary the time needed to reach the "Plateau", as well as its duration (LAKTRON, 38 p.).

2.6.4.4 Interpretation of   the curve

Figure 07 shows the temperatures measured by the sensor inside the test tube during an analysis.

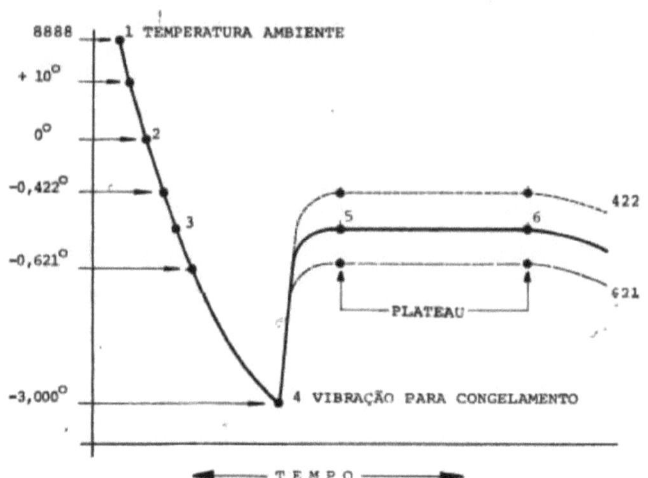

Figure 07: Temperatures measured by the sensor. SOURCE: (LAKTRON,

38

p.).

We have:

Point 1: Rapid cooling begins with the sensor and homogeniser already inside the sample during analysis.

Point 2: When you reach "point 2" on the curve, the temperature inside the test tube will be 0 H and at this point the counting starts, passing through points 422 and 621, until you reach point "4" which corresponds to a temperature of -3,000 H.

Point 3: Freezing point of pure milk.

Point 4: At this point the sample receives a violent vibration from the homogeniser lasting 1 second, which freezes the sample, starting the countdown by releasing the heat of fusion.

Point 5: The Plateau begins.

Point 6: The interval between points "5" and "6" corresponds to the time when the temperature remains stable, i.e. the "Plateau". When the point is reached

"6" stops the analysis and the result is shown on the device's display.

2.6.5 Fat

Compared to other fats, the fat in milk is a rich source of energy and serves as a transport medium for fat-soluble vitamins: A, D, E and K. It is found in emulsion form. Milk fat, which has an average value of 3.9%, is directly linked to various physical, chemical and sensory characteristics (DIAS, 2010).
The most commonly used method for determining fat in milk is the Gerber method, which is based on breaking the milk emulsion by adding sulphuric acid and isoamyl alcohol, centrifuging and then determining the fat. This determination can also be made using automatic apparatus (INSTITUIÇÃO ADOLFO LUTZ, 1985, p. 207-208).

CHAPTER 3

MATERIALS AND METHODS

3.1 MATERIALS

Five different samples of pasteurised whole milk types A, B and C sold in the city of Assis-SP were used.

3.1.1 Glassware

Burette 10 mL

Milk butyrometer

125 mL Erlenmeyer flask

250 mL beaker

Test tube

Test tube holder

11 mL volumetric pipette

10 mL volumetric pipette

1 mL volumetric pipette

100 mL beaker

1000 mL beaker

Glass funnel

3.1.2 Equipment

Bain Marie (TECNAL - TE-054).

Microprocessor Digital Cryoscope (LAKTRON - M. 90/BR)

Centrifuge (EXCELSA BABY I - MOD. 206)

Gerber centrifuge (FANEM - MOD. 202)

Thermolactodensí metre

pH meter (MS TECNOPON - MPA 210)

3.1.3 Reagents

The reagents used to prepare the solutions were of analytical grade.

Sulphuric acid P.A

Iso-amyl alcohol

Sulphuric acid Density 1.820 g/ml

Sodium hydroxide 0.11 M

Standard Solution A

Standard Solution B

Phenolphthalein

Buffer solution pH 4.0

Buffer solution pH 7.0

3.2 METHODOLOGY

3.2.1 Determination of Acidity in Degrees Dornic

The method used to determine acidity in Degrees Dornic was according to (INSTITUTO ADOLFO LUTZ, 1985, p. 203 - 204).

Procedure: 10 mL of the sample was transferred to a 100 mL beaker using a volumetric pipette. 5 drops of phenolphthalein solution were added. It was titrated with N/9 sodium hydroxide solution using a 10 mL burette until a pink colour appeared. The reading was taken and the result given in degrees Dornic.

Note: each 0.1 mL of N/9 sodium hydroxide solution is equivalent to 1°D.

3.2. 2Determination of Water Addition by Electronic Cryoscopy

Each device has its own specifications, so we followed the method and instructions of the manufacturer of the Laktron device (LAKTRON, 24 p.). which was used in the analyses.

A cooling bath solution was prepared in a 1000 mL beaker containing 150 mL of distilled water, mixed with 150 mL of glycerine and then with 300 mL of ethyl alcohol and shaken well.

The cooling chamber was filled with the solution from the cooling bath using a funnel until the level was reached, which could be seen on the right-hand side next to the drain tap. It is recommended that the solution from the reservoir be added daily, maintaining the recommended level. The reservoir should be drained every 20 days and a new solution added.

In order to obtain a good analysis, a test tube suitable for determining cryoscopy was used, as the tube needs to be able to cool down and contain exactly 2.5 mL of milk or standard solution inside.

The device was calibrated using 2.5 mL of standard solution A (0.422 °H), placing the test tube in the cooling chamber under the head. The Operation key was pressed. The head with the sensor was lowered and then calibrated to 0.422 with a tolerance of plus or minus 0.002°C or 0.002°H (Hortvet), depending on the device's specifications, and then calibrated using 2.5 mL of standard solution B (-0.621 °H) and the Operation button was pressed. After calibrating the device, 2.5 mL of the milk sample was added and the Operation button was pressed, waiting for it to stabilise and the result was recorded. After each reading, the sensor was carefully rinsed with distilled water and dried with absorbent paper.

3.1.4 Density determination at 15° C

The methodology used was according to (INSTITUTO ADOLFO LUTZ, 1985,

p. 199 - 200).

Procedure: 250 mL of the previously homogenised and cooled milk sample was transferred into a 250 mL beaker to allow the thermolactodensimeter to be inserted. The temperature was varied from (10-20)°C. The thermolactodensimeter was introduced slowly, avoiding immersing it beyond the point of emergence and taking care not to touch the walls of the cylinder. Wait for the mercury column of the thermometer and the densimeter to stabilise. The density and temperature were read.

Density at 15°C was expressed using a table for correcting density from temperature. The lactodensimetric degree values correspond to 2^a, 3^a and 4^a decimal places of the density value. To obtain the corrected density value at 15°C, simply place 1 to the left of the lactodensimetric degree value obtained from the table.

3.1.5 Determination of lacto sedimentation

The method is based on the precipitation of cells and dirt that may be in the milk, making it possible to check for undesirable impurities such as dirt, blood, faeces or other impurities.

Procedure: 10 mL of the milk sample was transferred to a test tube using a 10 mL volumetric pipette. It was placed in a centrifuge and left to run for 5 minutes. After the time had elapsed, the tube was removed from the centrifuge and it was observed whether any cells or dirt had formed at the bottom. The result was expressed as good, fair or poor.

3.1.6 Determining pH

The method used to determine the pH of the milk sample was according to (INSTITUTO ADOLFO LUTZ, 1985, p. 27).

Procedure: The pH meter was calibrated with pH 4.0 and 7.0 buffer solutions. Then 50 mL of the previously homogenised milk sample was added to a 100 mL beaker. The electrode was inserted into the sample and the pH value was allowed to stabilise and recorded.

3.1.7 Determination of Fat by the Gerber Method

The method used to determine fat was the Gerber method (INSTITUTO ADOLFO LUTZ, 1985, p. 207 - 208).

Procedure: 10 mL of sulphuric acid with a density of 1.820 g/cm^3 was transferred to the butyrometer using a volumetric pipette. Using a volumetric pipette, 11 mL of the sample was added slowly, making sure that it didn't burn on contact with the acid. Then 1 mL of isoamyl alcohol was added using a volumetric pipette.

The neck of the butyrometer was wiped with absorbent paper and the butyrometer was corked and shaken until completely dissolved. The butyrometer was centrifuged at 1200 +/- 100 rpm for 5 minutes, then placed in a water bath at (65 +/- 2) °C for 2 to 3 minutes with the stopper down. The stopper was handled, placing the light yellow, transparent layer (fat) inside the graduated scale of the lactobutyrometer. The value obtained on the scale corresponded directly to the percentage of fat, which was read on the lower meniscus.

CHAPTER 4

RESULTS AND DISCUSSIONS

In the tables below, the values that are not permitted by Normative Instruction no. 51 of 18/09/2002 are indicated in bold, with the exception of the determination of fat which, as it is whole pasteurised milk, the percentage of fat is in accordance with the original content, so we cannot assign a standard value. This is to make it easier to understand which samples could not be marketed.

Table 11 shows the results obtained from the five batches of sample A, with their respective standard.

Analysis	Legislation	Lot 1	Lot 2	Lot 3	Lot 4	Lot 5
PH	6,6-6,8	6,61	6,65	6,74	6,76	6,60
Lactosed i mentation	good	good	good	good	good	good
Acidity (°Dornic)	14-18 °D	15°D	14,5°D	15°D	15°D	16°D
Density (g/cm)³	1,027-1,034	**1,026**	1,029	1,028	1,0296	1.0312
Fat (%)	3,0%	3,3%	3,4%	3,9%	3,5%	3,5%
Cryoscopy (°H / %)	Maximum - 0,530/2,0%	**- 0,489 / 10,8%**	**- 0,514 / 5,0%**	**- 0,513 / 5,4%**	- 0,536 / 0,8%	- 0,555 / 0%

Table 11: Results obtained from sample A.

Analyses of pH, lactosedimentation and acidity in °Dornic of sample A showed all batches to be within the standard laid down by legislation.

In terms of density, the only batch that didn't indicate a value within the permitted range was number 01, which obtained 1.026 g/cm³ , and the minimum permitted value is 1.027 g/cm³ .

In the digital cryoscopy analysis of the first three batches, the results obtained were disapproved as they were above 5.0% water, and the permitted level is up to 2.0%. Batches 04 and 05, on the other hand, showed results of 0.8 and 0 per cent water respectively, which are considered correct by the legislation. An interesting fact was the decrease in density in batch 01, as the more water added to the milk, the lower the density, and this batch was the one with the highest value of all, i.e. 10.8 per cent water.

Table 12 shows the results obtained from the five batches of sample B, with their respective standard.

Analysis	Legislation	Lot 1	Lot 2	Lot 3	Lot 4	Lot 5
PH	6,6-6,8	6,61	6,69	6,82	**6,93**	6,72
Lactosed i mentation	good	good	good	good	good	good
Acidity (°Dornic)	14-18 °D	18°D	15°D	14°D	**12°D**	15°D
Density (g/cm)^3	1,027-1,034	1,0289	1,0294	1,0274	**1,0256**	1.0294
Fat (%)	3,0%	2,7%	3,4%	2,9%	2,4%	2,8%
Cryoscopy (°H / %)	Maximum - 0,530/2,0%	- 0,558 / 0%	- 0,544 / 0%	**- 0,529 / 2,2%**	**- 0,528 / 2,4%**	- 0,562 / 0%

Table 12: Results obtained from sample B.

The pH determination of sample B showed that only batch 04 was outside the standard set by the legislation, and consequently its acidity was 12°D, justifying the high pH value of 6.93.

Lactosedimentation was good in all five batches.

The density of batch 04 sample B (1.0256 g/cm^3) was the only one below the minimum allowed by law (1.027 g/cm^3).

Of all the samples, this was the one with the best digital cryoscopy values. Batches 01, 02 and 05 had 0% water in the milk, and only batches 03 and 04 had 2.2 and 2.4% water respectively, i.e. only 0.2 and 0.4% water above the legal limit.

The sample from batch 04 had a pH of 6.76 and an acidity of 12°D. This could mean that the milk is alkaline, which could mean that it is milk from a cow with mastitis, milk from the end of lactation, retention milk or milk that has been defrauded with water.

Table 13 shows the results obtained from the five batches of sample C, with their respective standard.

Analysis	Legislation	Lot 1	Lot 2	Lot 3	Lot 4	Lot 5
PH	6,6-6,8	6,68	6,70	6,82	6,76	6,71
Lactosed i mentation	good	**regular**	**regular**	good	**regular**	**regular**
Acidity ("Dornic)	14-18 °D	16°D	16°D	15°D	14°D	16°D
Density (g/cm)^3	1,027-1,034	1,0276	**1,0264**	1,0286	1,0276	1.0284
Fat (%)	3,0%	3,3%	2,8%	3,3%	2,9%	3,2%
Cryoscopy (°H / %)	Maximum - 0,530/2,0%	**- 0,482 / 11,6%**	**- 0,447 / 18,6%**	**- 0,517 / 4,6%**	**- 0,473 / 13,4%**	**- 0,501 / 7,8%**

Table 13: Results obtained from sample C.

The pH analysis did not produce satisfactory results, with only two batches, 03 and 04 (6.82 and 6.69, respectively) indicating values within the legal limits. Lactosedimentation also showed an unsatisfactory result, the only exception being batch 03, which obtained a good result, implying that there is dirt in the milk, and that this type of milk could not be going to the public for consumption. The acidity in °Dornic and the pH are in line with legal standards in all five batches of milk.

In terms of density, only batch 02 (1.0264 g/cm^3) is not within the legal limit (1.027 g/cm^3).
All the batches in sample C failed the digital cryoscopy test, showing high levels of water in the milk, particularly batches 01, 02 and 04, which showed a percentage of added water above 10.0% in the milk, especially batch 02, which obtained 18.6%, and of all the cryoscopy analyses carried out on the five samples from five different batches of milk, this was the one with the highest level of added water.

Table 14 shows the results obtained from the five batches of sample D, with their respective standard.

Analysis	Legislation	Lot 1	Lot 2	Lot 3	Lot 4	Lot 5
PH	6,6-6,8	6,66	6,70	6,72	6,75	6,67
Lactosed i mentation	good	good	good	good	good	good
Acidity (°Dornic)	14-18 °D	15°D	15°D	15°D	14°D	15°D
Density (g/cm)^3	1,027-1,034	1,0279	1,0286	1,0276	1,028	1.0294
Fat (%)	3,0%	2,7%	2,9%	2,6%	2,9%	2,7%
Cryoscopy (°H / %)	Maximum - 0,530/2,0%	**- 0,503 / 3,4%**	- 0,531 / 1,8%	- 0,530 / 2,0%	**- 0,527 / 2,6%**	- 0,563 / 0%

Table 14: Results obtained from sample D.

The pH of the samples from batches 01, 02 and 05 (6.57, 6.50 and 6.39 respectively) is below the minimum permitted pH value of 6.60. Only batches 03 and 04 (6.72 and 6.70 respectively) would be approved by Brazilian legislation.

The pH, °Dornic acidity, lactosedimentation and density determinations were all within the standard.

When it came to determining the fat content of the milk, all the batches registered a percentage below 3.0 per cent.

The cryoscopy carried out on batches 01 and 04 (3.4% and 2.6% respectively) were the only ones to show a percentage of water above the legal limit.

Table 15 shows the results obtained from the five batches of sample E, with their respective standard.

Analysis	Legislation	Lot 1	Lot 2	Lot 3	Lot 4	Lot 5
PH	6,6-6,8	6,60	**6,28**	6,78	6,72	6,62
Lactosed i mentation	good	good	good	**regular**	**regular**	**regular**
Acidity (°Dornic)	14-18 °D	15°D	**20° D**	14°D	15°D	18°D
Density (g/cm)^3	1,027-1,034	1,0292	1,0299	1,0276	**1,0268**	1.0296
Fat (%)	3,0%	3,2%	3,4%	3,3%	**3,9%**	3,4%
Cryoscopy (°H / %)	Maximum - 0,530/2,0%	**- 0,502 / 7,6%**	**- 0,516 / 4,8%**	**- 0,483 / 11,4%**	**- 0,502 / 7,6%**	**- 0,520 / 4,0%**

Table 15: Results obtained from sample E.

Batch 02 failed the pH test, as it had a pH of 6.28, which is below the minimum 6.60 that a milk sample should have. As a result, its acidity also exceeded what is permitted by law, which is a maximum of 18°D.

Batches 03, 04 and 05 showed little dirt in the milk in the lactosedimentation test, which makes it unfit for consumption.

In determining density, the only batch that did not show a value within the permitted range was number 04, which obtained 1.0268 g/cm^3 , and the minimum permitted value is 1.027 g/cm^3 .

When it came to determining fat, all the batches showed a percentage above 3.0 per cent, with the highlight being batch 04, which showed the highest percentage of fat of the entire study, 3.9 per cent.

In the cryoscopic analysis, all the batches in sample E were not in line with the legislation, as they all showed values above the double permitted amount of water in the milk. Of particular note was batch 03, which showed 11.4 per cent water, a value considered high for a sample.

The sample from batch 02 had a pH of 6.28 and an acidity of 20°D. From this data we can conclude that the milk was slightly acidic, at the beginning of lactation with colostrum and at the start of the fermentation process.

Despite these results, we have to take into account that the percentage of water in milk varies according to the way the milk is stored and its hygienic and sanitary condition. It also depends on the animal's diet, the seasons, i.e. the period in which milking took place, and finally the breed of animal. But even so, it is understood that the 2.0 per cent of water allowed by law in the milk sample is precisely because of these factors.

Even with this information, we have to consider that some of the results obtained, such as batches 01 of sample A, batches 01, 02 and 04 of sample C and batch 03 of sample E, indicated a percentage above 10.0%, which means that water may have been added to these samples, especially batch 02 of sample C, which showed 18.60% water.

Figure 08 shows the average values of the results obtained.

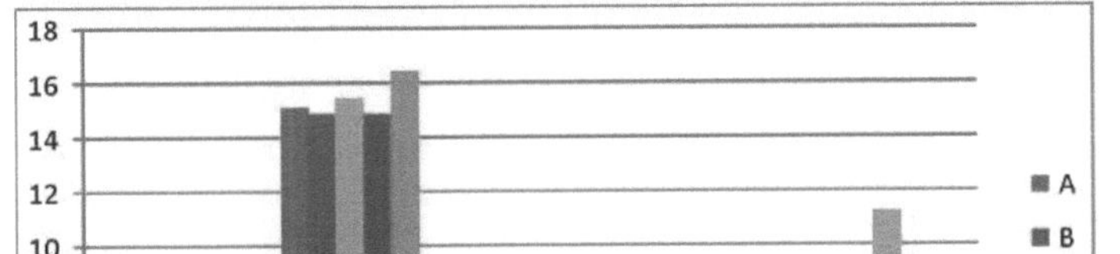
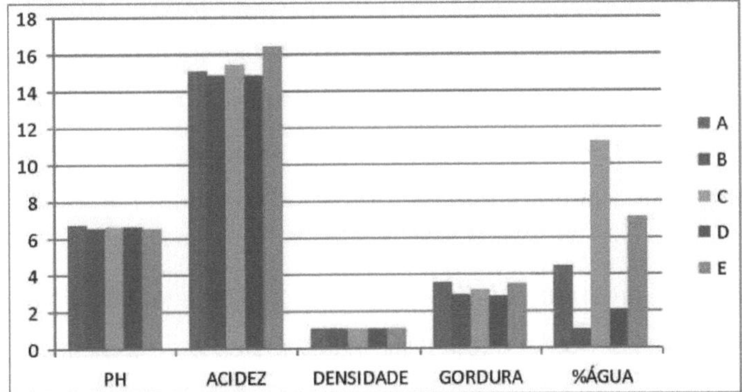

Figure 08: Average values of the results obtained.

This graph shows that brands A, C and E have added more water than permitted.

35

CHAPTER 5

CONCLUSION

The determination of acidity in the milk samples was considerably positive, as only batch 04 of sample B and batch 02 of sample E showed values outside the standard set by the legislation.

On the other hand, several pH values were not up to standard, of which two batches were from sample E, and three batches each from samples B, C and D, coincidentally all from batches 01, 02 and 05.

The density analysis also obtained satisfactory results, with only four of the 25 batches showing values below or above those permitted by legislation.

From what we have seen, we can see that the main negative point was that of the 25 batches of milk used to carry out this work, only three (12%) have all the parameters in line with Normative Instruction No. 51 of 18/09/2002, and are therefore suitable for consumption by the population of the city of Assis-SP.

Another important point was the digital cryoscopy analysis, which showed that all batches of samples C and E had values above 2.0% of added water in pasteurised milk, which is the maximum permitted by Brazilian legislation.

When determining lactosedimentation in the milk of sample C, only batch 03 obtained a result with no dirt in the milk, all the others indicated the presence of small dirt in the milk, which already makes it unfit for consumption by the population.

Sample E also showed the presence of dirt in batches 03, 04 and 05, which already indicates a 60% occurrence of dirt in batches collected over a 5-month period in 2013.

Due to lack of time, microbiological tests could not be carried out on these samples, which could indicate the presence of micro-organisms.

CHAPTER 6

BIBLIOGRAPHICAL REFERENCES

BECCHI, Cleusa Scapani. **Study of the cryoscopic index of "in natura" type B milk produced in the dairy basin of the Taquari Valley, RS**. 2003. 106 p. Dissertation (master's degree). Postgraduate Programme in Veterinary Sciences - Federal University of Rio Grande do Sul, Porto Alegre, 2003.

BEHMER, Manuel Lecy Arruda. **Milk Technology**. 13 ed. São Paulo: Nobel, 1999. 320 p.

BEHMER, Manuel L. A. **How to make good use of your milk on the farm**. 6. ed. São Paulo, Editora SP: Nobel S.A, 1984.

BEUX, Simone. **Dairy technology workbook.** Federal Technological University of Paraná - Pato Branco Campus. Available at: < > Accessed on 11 October 2012.

BRAZIL. Decree No. 30.691, of 29 March 1952. **Approves the New Regulations for the Industrial and Sanitary Inspection of Products of Animal Origin.**

BRAZIL. Normative Instruction 51, 18 Sep. 2002, Revokes Ordinance no. 146, 7 Mar. 1996.
Technical regulations for the identity and quality of dairy products. Federal Official Gazette, Brasilia, 20 September 2002.

BRITO, Maria Aparecida; ARCURI, Edna; LANGE, Carla; SILVA, Márcio; SOUZA, Guilherme; BRITO, José Renaldi. **Titratable acidity.** EMBRAPA Information Agency - Milk Agribusiness. 2006. Available at: <http://www.agencia.cnptia.embrapa.br/Agencia8/AG01/arvore/AG01_194_2 1720039246.html> Accessed on: 15 October 2012.

DIAS, Ana Maria Costa. **Analyses for milk quality control.** 2010. 42 p. Technological Specialisation Course in Food Quality. Polytechnic Institute of

Coimbra - School of Agriculture, Coimbra, 2010.

FONSECA, L.F.L.; SANTOS, M.V. Estratégia **para controle de mastite e melhoria para a qualidade do leite**. Barueri, SP: Ed. Manole; 2007. 314p.

GUIMARÃES, Pautilha. **Milk Science. Pasteurised Milk**, Minas Gerais, 2001 available at: <http://www.cienciadoleite.com.br/leitepasteurizado.htm> Accessed on 11 October 2012.

ADOLFO LUTZ INSTITUTE. **Normas Analíticas do Instituto Adolfo Lutz**. v. 1: Métodos químicos e físicos para análise de alimentos, 3. ed. São Paulo: IMESP, 1985, p. 203-204.

INSTITUTO ADOLFO LUTZ **Normas Analíticas do Instituto Adolfo Lutz**. v. 1:
Chemical and physical methods for analysing food, 3. ed. São Paulo: IMESP, 1985, p. 199-200.

ADOLFO LUTZ INSTITUTE. **Normas Analíticas do Instituto Adolfo Lutz**. v. 1, Métodos químicos e físicos para análise de alimentos, 3. ed. Sao Paulo: IMESP, 1985. p. 27.

ADOLFO LUTZ INSTITUTE. **Normas Analíticas do Instituto Adolfo Lutz**. v. 1, Métodos químicos e físicos para análise de alimentos, 3. ed. Sao Paulo: IMESP, 1985, p. 207-208.

BRAZILIAN INSTITUTE OF GEOGRAPHY AND STATISTICS (IBGE). **Municipal Livestock Survey, 2011.** Available at: <ftp://ftp.ibge.gov.br/Producao_Pecuaria/Producao_da_Pecuaria_Municipal/2 011/ppm2011.pdf> Accessed on: 22/03/2013.

LAKTROM Indústria e Comércio de aparelhos eletrónicos LTDA. **M-90 Microprocessor Digital Cryoscope Operating Manual.** 38 p.

LISBÔA, Júlio Cezar Foschini; BOSSOLANI, Monique. Dairy experiments. **QUÍMICA NOVA NA ESCOLA magazine,** No. 6, November 1997, p. 31.

PAIVA, Regina Márcia Bahia. **Physico-chemical and Microbiological Evaluation of Type C Pasteurised Milk Distributed in a Government Social Programme**. 2007. p. 76. Dissertation (Master's) - Department of Veterinary Medicine - Federal University of Minas Gerais, Belo Horizonte, 2007.

ROCHA, Luciana Sampaio Valões da. **Evaluation of the physical, chemical and microbiological characteristics of Type C pasteurised milk sold in the municipality of Maceió -AL**. 2009. 30 p. Final Coursework. Specialisation in Hygiene and Inspection of Products of Animal Origin - Universidade Federal Rural do Semi-Árido - UFERSA, Pernambuco, Recife, 2009.

SCALCO, Tiago. Milk in a plastic bag vs. milk in a long-life carton. **Sul do Brasil Rural**, Chapecó. 29, November. 2009. p. 01.

SILVA, Paulo Henrique da. Composition and Property Aspects. **QUÍMICA NOVA NA ESCOLA magazine**, No. 6, November 1997, p. 01.

SOUZA, Loiane Mayra Jacó de. **Veterinary Medicine Course Final Paper**. 2006. 66 p. Course Conclusion Work. Department of Veterinary Medicine - UPIS Faculdades Integradas, Distrito Federal, Brasília, 2006.

TRONCO, Vânia Maria. **Manual for Milk Quality Inspection**, 2. Ed. Santa Maria: Editora UFSM, 2003. 192 p.

TURCO, Cristiane de Paula; COSTA, Guilherme Machado; PAIVA, Hélio Afonso Braga; BARROSO, Marcelo Francini Girão; CÔNSOLI, Matheus Alberto; NOGUEIRA, Maurício Palma; ROSSI, Ricardo Messias; SILVA, Rosana de Oliveira Pithan; TORGGLER, Sérgio Pinheiro; NETO, Sisgismundo Bialoskorski. **Planejamento e Gestão Estratégica do Sistema Agroindustrial do Leite no Estado de São Paulo**, 1. ed. Translation: SEBRAE, São Paulo, Editora Atlas, 2007.

Printed by Books on Demand GmbH, Norderstedt / Germany